The Gamecock

GAMECOCK

The Quality Magazine Devoted To Game Fowl

SINGLE COPY $8.00   JULY 2015   U.S.A. $28.00 YEAR

HACKLE

On T it I   ...ommy Carrano In New York

Game fowl and the Foul truth True story of Tommy & Gina Carrano

AuthorHouse™
1663 Liberty Drive
Bloomington, IN 47403
www.authorhouse.com
Phone: 833-262-8899

This book is printed on acid-free paper.

ISBN: 979-8-8230-1373-4 (sc)
ISBN: 979-8-8230-1374-1 (hc)
ISBN: 979-8-8230-1372-7 (e)

Library of Congress Control Number: 2023916144

Print information available on the last page.

Published by AuthorHouse 08/25/2023

authorHOUSE®

GAME FOWL
AND
THE FOUL TRUTH

True Story of Tommy & Gina Carrano

Introduction

My name is Thomas Carrano. If you've heard my name, you know my story. What follows is my plight—the painful details, the infuriating trauma, and the heartbreaking conviction. All of it. I need to tell this story, because it needs to be told, not simply for my own benefit—although that's part of it—but because what happened to me could happen to anybody. This story is for all of us.

If this were a movie, the scene I'm about to describe would be the height of the action. The climax. It still plays in my mind like an old film. It blurs, speckles with time, and maybe cuts out once or twice, but the picture is there. It's all still there.

In May of 2017, my wife, Gina, and I had our farm raided by the Bronx police, the US Department of Agriculture, and the ASPCA. It was six in the morning, and I was with my wife in the kitchen. Gina and I had moved to the country of western New York nine years ago, both having lived in the downstate area for most of our lives. We moved to be in a good place and to preserve what amounted to my life's work: raising game fowl. It was a different pace of life than we were used to, but we had adjusted and made it our home. We enjoyed our life.

Gina's one of the most kind-hearted people you'd ever want to know. She's a special education teacher, former Sunday school teacher, and a great wife—there's a reason her nickname is Snow White.

That morning wasn't unusual for us. We sipped coffee in the dark hours of dawn while listening to an inspirational program and getting ready for the work ahead of us. Gina and I appreciated the stillness of early mornings. That time was clarifying, set against the daily trials we would have to face.

Coffee still in our hands and thoughts reflective, we heard a bang at our front door. Then another one. Loud, successive bangs on our front door at six in the morning. Four federal agents stood on our front step, head to toe in gear and braced for battle. Next thing I knew, they were barging inside my home. There were noises—ranting shouts that were mostly inaudible. One demanded to know where I kept my guns; the others' voices clanged together, like echoes of a barroom brawl.

Gina, bewildered, thought someone was hiding on our property. Despite our shock, we wanted to help sort out the chaos; that was until we saw no less than forty ASPCA workers huddled outside. Behind them were several box trucks and crime tape around the house. All the way from New York City, the trucks sat, hunched on our front lawn. Their white paint shining in the sunrise. All those guests came all that way, to our hobby farm in the country, to say good morning.

That morning started it all.

1

GAME FOWL AND THE FOUL TRUTH

There was the raid, and there was the aftermath. I couldn't tell you which was worse to live through. Now here I am, sitting in McKean Federal Correctional Institution's finest accommodations and doing time for a crime I didn't commit. That's the elusive term for me at the moment—time. Can't define it; couldn't begin to quantify it. I know it's early in the day, not from the bleak rays of light illuminating the crevices of my cell but because I woke to the sound of a skull getting caved in a few cells down. See, when you're locked up, time presents itself in earnest. It's up to you to decide its distinction as friend or foe. Believe me, that's a crucial decision to make in here. Time won't go anywhere. All I know is that I have too much of it right now.

What I can do is write. Reliving this stuff isn't easy. Reflecting on what happened is hard enough, but knowing how wrong all of it was and how powerless I was to the whims of an abusive system … It's a lot to process when you're wearing the same jumpsuit you've seen in movies. But everyone's innocent in here, right?

Believe it or not, I'm not a communist. I had respect for this country's government and for the laws created to keep us safe. I believed what all American citizen are taught to believe—that the government is in service to its people. That it fights for justice. Many of you reading this probably honor those beliefs too. I'm not writing this stuff to trick you. Like any piece of literature, you take away what resonates with you. But this isn't fiction. What follows in this story happened. You decide if "justice" has the same meaning to you. I've got time.

Let's meet the characters of this saga. I'll do my best to introduce them with the least amount of provocation. Good stories have protagonists and villains, and good authors are bound by literary honor to obscure their intentions until their actions prove otherwise. But hey, I'm just a guy.

You may know me, but allow me to give a little more context. I own a farm that's registered on a state and federal level. Gina and I breed Old English Game fowl, something I've done since I was a kid. If you drive past our property, you'll see a sign that says, "TNT Game Fowl," before a many acre sprawl. That's us. It's my life's work, and I'm proud of it. Raising a rare breed of game fowl isn't easy. It takes dedication, hard work, and a dogmatic respect for the animals to make sure they're safe and healthy. It's a rich tradition, and I consider myself a student of its history. Keep that in mind. Credible Game Breeder's Association poultry shows show my game fowl. Sanctioned poultry judges judge them on confirmation, colors, and breeds.

Enter the ASPCA. Their mission is to rescue, protect, and care for animals in need through a wide range of activities, like animal relocation, advocacy, training, legislature, and veterinary services. They take massive amounts of money from thousands of nationwide donors—as of 2019, $280 million—to carry out their nominal pledge. You can do your own research to see where that money goes. The point is that they have money and influence—a lot of it. Keep that in mind.

Working in tandem with the ASPCA is the federal government. There's not much I can tell you about its power as an entity that you don't already know, but if it has you in its crosshairs, it grows as many slithering tentacles into your life as it needs. Whoever feeds it reaps the bounties of its access. Keep that in mind.

Like any good legal story, there are lawyers too. Some good lawyers, and some bad. Some want to help people like me, and some want copious amounts of money from people like me. Keep that … Well, you get it.

With that, let's resume.

2

THE CRUELTY OF HOPE MISPLACED

After the raid had commenced and the panic and confusion of what had happened to my quiet little farm was just beginning to settle, I realized that I was going to need help. In more concrete terms, I needed a lawyer. Given my profession, I was about as green as they came in regards to navigating complicated legal advice. I was desperate to understand what was happening and what I was up against. Looking back, I couldn't have been an easier target for the sharks swimming laps around my despair. I met with a firm, who told me what I wanted to hear—that there was no case and that this was all intimidation and scare tactics—for five grand a week.

As helpless as the raid had made me feel, I bought into the illusion of hope these people offered. It was a rich oasis in the middle of an oppressive, suffocating desert. Boasting the hundreds of cases under their belt, I put my trust and money in these people. After meeting with a dozen lawyers, I didn't really have much of a choice. Gina and I had worked hard over the years to build a life for ourselves, but we were not wealthy people. I'm a farmer, and she's a teacher. That firm was who we could barely afford, even with help. It's hard to know exactly what and who to trust when you're bruised, battered, and drowning in fear. For these predators in disguise, that fear and insecurity might as well have been dollar signs tattooed on our foreheads. Stripped of dignity, all we had was hope that those people might be able to help us.

Put simply, they did not. As it turned out, they weren't even practicing federal lawyers. It was a classic bait-and-switch. Two weeks before the trial, the lawyer who I had been working with for an entire year—the man who was supposed to represent me—completely ditched the case after the first hearing. He sent a representative for the others, putting someone who couldn't tell you the first thing about game fowl in his place. The representative even thought game chickens were like salmon! This vanishing act was after I had already given $50,000 to this firm. The lawyer reappeared during the week of the actual trial to call me and ask for the balance of funds. After everything, they just wanted more money. The oasis, like the allegory suggests, was a mirage. I was alone in the desert, and there would be no water. The hope I bought into—quite literally—was drying up.

I prayed for guidance and direction. On a visit down to Long Island, I stopped by the house of my sister-in-law Barbara. We sought our faith for answers in the face of this hopelessness. Out of those prayers came the name for a book, thanks to a friend of Barbara's. *Game Fowl and the Foul Truth*. I thought it was perfect. It described my situation. While the trial loomed, I got busy writing, and within three months, we had finished our first book. Gina and others were of significant help. I started that book not just with the intention of raising money for our bills and outrageous lawyer fees but, more importantly, to bring education to the game fowl community and other people who were not familiar with the old breeds of game fowl.

Thankfully, the book sold well. With that and with help from Gina's family, the Vaccaros, who sold equipment on the farm, I retained a new lawyer, Richard Willstatter. I'll name him, because he deserves to be named. He was a whiz with federal law, and he'd fought many battles for people like me. He wouldn't let me believe a delusion. Richard knew immediately what we were going to be up against. He also recognized the giant hole that the blunders my previous incompetent representation had put us in. Our prayers were answered for the time being. But would it be too late?

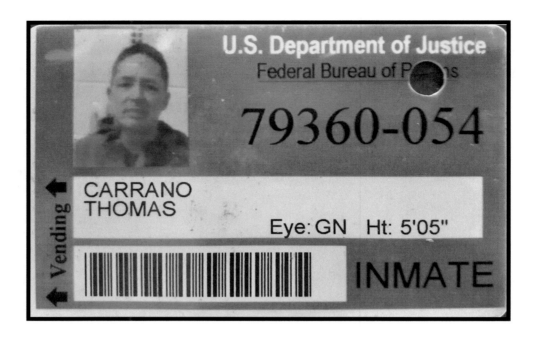

3

THE TRIAL

N ow we're in New York City, the setting for my federal trial. Notice anything peculiar? New York City is known for a lot of facets of modern life, but complicated agricultural procedure is not one of them. I'd wager many of the jury—a few of them ASPCA donors—didn't have much experience with chickens, other than knowing how they were cooked and served in their favorite French bistros. It was a strategic move, holding this trial far away from communities that understood agriculture— nearly three hundred miles from my farm. They shuffled into their benches on the first day of a five-day trial, ready for a freak show. I, the fearsome upstate farmer, was the main attraction. They weren't disappointed.

Finally, it was time for the great unveiling. The prosecution showed their hand. You'd think that for all this to happen, they'd have to have some damning evidence. Otherwise, all the plea deals they were offering me before the trial could have been considered gigantic bluffs!

I don't even know where to start. Watching that mockery of a trial unfold right in front of me is something I'll never forget. Sometimes I can't help but laugh, thinking back to it. Other times, I'm left with profound sadness at the state of our justice system. I'll try to put this as plainly as possible: they didn't have jack.

While the raid was happening, I knew they were confiscating a great number of items from around my house. Those turned out to be a collection of historic memorabilia—history books about cockfighting, gaffs (three-inch long picks for the spurs), and more. Presented to a jury without context or knowledge of raising game fowl, I knew immediately what the prosecution was trying to do. They read passages from those history books for the shock value, but the only connection they could make was that they were under my possession. The prosecution blew up pictures of the gaffs I had collected on the big screen to make them look as menacing as possible. Again, they weren't trying to prove that I actively used them, because they couldn't! The prosecution was attempting to establish a narrative to a naive audience. It was working.

Let's talk about Facebook for a moment. If you have a Facebook account and use it actively to communicate, delete it. Take it from me. There were a few search warrants issued by the federal government at my expense. One allowed them to ask Facebook for all of my data—anything I had posted publicly, of course, but also private messages sent using their platform. There were messages sent to other members of the game fowl community that the prosecution added to their narrative. Context matters in private conversations, but pages and pages of messages that would've added that crucial element were redacted, leaving a few short phrases to individuals that were not relevant to anything other than the narrative they were constructing. I don't know how this was even allowed to be entered as evidence. Maybe if I'd had better representation sooner, we could've easily pointed out the blatant obstruction of truth in their claims. Again, no evidence directly linked me to anything that they were trying to claim. Hindsight, right? I sat there and took it, knowing that I was fighting a losing battle.

Next came the "expert" witness the prosecution had brought in. This guy should've been an actor. He was, allegedly, an undercover law enforcement official, whose sole purpose was to infiltrate cockfighting rings in various parts of the United States. He didn't have any experience in New York, however. As far as I know, this was a free trip to the big apple! But he was a very good storyteller and cast a spell over the jury about the cruelty and abuse of game fowl he had witnessed in his profession. I don't doubt his experience in that regard. But did it have anything to do with me? Of course not. There wasn't even a mention of me. That's not what the prosecution was going for. Remember, it was a show. The vivid stories he was sharing went straight to the heart of the jury members. He didn't need to have any connection to me to be effective. However, the "expert" witness lacked any knowledge of raising game fowl. He made a few blatant statements that had no basis in fact. He was, essentially, an actor playing a role, and he played it well.

4

The trial ended, and I got my time. I said goodbye to my beautiful wife and extended family and headed to prison. Divorced from the drama of everything, there are some very real consequences I'm battling through. This last section might be the most challenging to write about, so bear with me.

Unless you've been there yourself, it's difficult to express how severe a toll all these false accusations have on your mental being. The dark, rotting hopelessness creeps into your brain and makes itself home. You feel despair daily, knowing that you're innocent, but you're unable to fight against the most powerful entity in the United States, which wants nothing more than your destruction. And for what? What was all this for? Is that really the country we live in? It's even worse knowing that this despair is exactly what was intended.

Once you're accused of something, you really see the true colors of those you thought you knew. This book could be a broader discussion about cancel culture. That's a very sick part of our society that doesn't want to look for the truth anymore; it's just easier to react to the drama. I lost a lot of respect for people and organizations that I used to work closely with. All it took was the word of what had happened to me, and relationships that I had spent years of my life building vanished without a trace. It's like you're infected with a corrosive disease. No one wants to touch you. Unless you've been there yourself, it's hard to describe the loneliness and disappointment. The American courage, which used to be a national pride, is eroding past the point of no return. We're all marching toward a cruel, isolated future, and it makes me very sad.

That said, I wouldn't be here without help from those around me. I touched on this a bit in my first book, but there is power in community. The nationwide game fowl allies I've made, who stood by me during this ordeal, have my loyalty for life. Kayce Chadborn started the Tommy Carrano Legal Defense Fund, which was a huge part of being able to get competent representation. Without the support that we had, fighting this would've left Gina and me in financial ruin. Gina's family showed me the collective strength of a family who cares.

Gina is an amazing woman. If we judge people on times of extreme adversity, then I couldn't be more proud of the woman I married. Through all the panic, fear, and helplessness, she was there for me, keeping me strong. When the plea deals started coming in, she was the first one to beat them away. She knew what I would have to sacrifice to cooperate with the prosecutor's scam. She taught me what true loyalty and selflessness means.

Our temporary separation hasn't been easy, but I believe love's strength derives from situations like these. We are united with God, walking hand in hand through his Word. Gina's strength keeps me going when I feel weak. Those trips to New York City for the trial procedures are painful memories, but what really shines through that darkness is the love I have for my wife. Her heart wouldn't let me sink. God blessed us with each other, and he will provide. Our time is coming.

That's my story.

Inspected by a NY State department of agricultural inspector when Pumpkin was returned to Gina while I was still in prison

Epilogue

When someone has been through all of this, there are things to take away and lessons to learn. It seems unfathomable to have to go through all that I did for chickens, but anything worthwhile in life should be fought for. After I returned from prison, we filed an appeal to no avail. They would not retry the case, because I had served the time. I'm a convicted felon over chickens. They only had hearsay evidence from

Facebook, but somehow they were convinced that I conspired to break the animal welfare act. They had no witnesses or wiretaps—really nothing to substantiate their claims. The New York City jury could not possibly have understood the type of birds involved. To say that this has been frustrating is an understatement. I put in for a presidential pardon with President Trump, and I will continue applying. No one should have to have a record over this.

My wife and I continue to take care of the remaining birds sent back from the "care" of the ASPCA. Many died. The ones that lived were in horrible condition. We will forever be grateful to our civil lawyer, Steven Kessler, who fought to have these birds returned to us.

I have since found a national organization for the preservation of game fowl. I am working with many people who have and are going through the same type of trial. It is my sincere hope that organizations who claim to protect animals do just that.

The cover of this book says it all. I was involved with the United Game Fowl Breeders Association and sent to federal prison for raising game fowl. President Trump took office the year the raid on my farm took place. I found this interesting. Our country is in more turmoil than most want to recognize. In my opinion, we are in trouble. We need a sovereign act of God and strong leadership willing to pull on the reigns of injustice. The phrase "if you do not stand for something, you will fall for anything" has been a quote that I have visited many times over the last couple of years. It has become clearer, because a plea deal was not something I would have ever lived with, not over something I had always enjoyed.

This whole case stemmed from when I was the President of the New York United Game Breeder's Association (NYUGBA). A member down in the Bronx said that he was taking me down with him over Facebook. I had never even met this member, yet I was sent to federal prison because of his comments. Can you imagine people being sent to prison on a regular basis over Facebook comments? What are we allowing if we let social media cases set legal precedence? We need perspective.

I started writing this book from a federal prison cell in the E unit, A range. Big shout out to the members of McKean FCI in Pennsylvania. It cost $40,000 of taxpayers' money. Guess what? I'm a taxpayer, and my case had no wiretaps, testimonies, or witnesses. Every government "expert" was from a different state. I do not believe that this is a direction our country wants to go in. It is borderline Nazi Germany. That is exactly how it has felt for my wife and me since May 23, 2017. I honestly cannot recount all the people I have spoken to who have been victims of little investigations and serious legal consequences. This ought not be.

We are moving forward and gaining momentum nationally. I believe that our books, *Game Fowl and the Foul Truth* and *One Count: The True Story of TNT Game Fowl Farm*, are going to spark national attention. God-fearing, tax-paying, law-abiding citizens should not be treated this way. Through the first book, we hoped to bring understanding that we must run our race. The race was the game fowl community and the agricultural community coming together and saying enough. Our rights need to be contended for, and the only way to do that is through education and unity.

When going through federal cases, having good lawyers and support is a must. I have become friendly with an older man by the name of Floyd Beadreau. He went through a similar situation over his dogs many years ago, in Louisiana. He was eventually found not guilty and gave me some of the best advice. He said, "When you are right, you are strong, and when you are wrong, you are weak!"

The day the government came to our little hobby farm, they wanted me to go to jail. The sad truth is that this is exactly what happened. They did nothing to investigate. They had pictures of my house gator, where chickens were dubbed, which is a common practice among breeders. That got them a search warrant. They also had Facebook pictures. I would post pictures of chickens from shows. They claimed that those chickens were my own. The whole case was built on a lie. The private messages had pages redacted. The messages they did share were missing context. One such message said, "He's liked fighting chickens his whole life."

They confiscated my phone, which had text messages. They used my phone to track my movements to the feed store. The prosecutor opened the trial by calling me the "national leader of the underground cock fighting." Because I had history books from people who fought chickens and kept my birds separate from roosters, I was charged with one-count of conspiracy against the animal welfare act

Not one of my neighbors was spoken to or sought out in any way. They did not even know that my farm was registered on a state and federal level. They were clueless to the fact that I had worked on preserving these fowl for an exceptionally long time. The fact that they had no idea how to care for these birds after they were confiscated was very disturbing!

The first inspection in September 2017, four months after my fowl were taken, was horrid! The birds were sickly, and many had died with no explanation. The government had tried to get the fowl on criminal forfeiture, but that was eventually got dismissed. It wasn't legal to take my birds. Why? Because a one count of conspiracy was not enough to take my fowl.

They were all in perfect health when confiscated. My fowl represented hundred-year-old lines that have been destroyed by too much governmental power.

The process of deterioration is heartbreaking. To date, twenty-five of my fowl have died, along with some hens that I had. Two of them, I'd had for sixteen years! They are long gone. The care that went into these birds on my farm was second to none. For my wife and me to see those once beautiful fowl destroyed by the hands of people who were supposed to prevent cruelty was devastating. They caused the cruelty! How this could ever be considered justice is beyond me. A miscarriage of justice would be a more appropriate description of my case.

They convicted me of breaking federal law 2156. I will continue to maintain my innocence. I have never conspired to break any laws. I did everything to follow the laws. I was hoping to have some of these laws changed in rural upstate New York. I could never have imagined that something I enjoyed doing my whole life would come to this.

My wife grew up in Long Island, but she had really adapted to owning and running the farm with me. We enjoyed it together. Our farm received compliments from everyone who came to visit. We had state inspectors give us stellar reviews every year. Unfortunately, the jury was made up of Manhattan people, who could not possibly understand what was being presented to them. They handed over a guilty verdict. Gina and I have lost so much since then. The finances to try to fight this almost cost us our homes. All over a chicken conspiracy to violate federal law 2156.

After my conviction, the government said that they would settle for half the jail time if I admitted to taking a leadership-enhancement position and waived my right to an appeal. I did not agree! I never had a leadership position, and I wanted to appeal. I could not agree to take a plea for something I never did. I had enjoyed game fowl for thirty-seven years. I could not lie just to get myself out of that God-awful situation. I had to continue to trust in God and the process of Justice, so prison it was.

The separation was difficult on Gina and me. Gina was left to manage the five acres by herself. She was OK. We knew that God was with us both and would use all of it for his purposes.

On December 13, 2018, I was sentenced in the southern district of New York. It was a nerve-racking experience. Gina and I got there the night before. The next day, the appointment was put off, from ten in the morning to noon, and then four in the afternoon. They wanted us to come back the next day. We could not, because Gina only had one day off. We lived seven hours away! Anyone who put themselves in our shoes would realize the hell of driving seven hours for every court date for the past year and a half.

I need people to understand that I was charged with one count of conspiracy to break the animal welfare act. This country consumes millions of pounds of chicken a year! In my case, there was no gambling, fighting venture, or drugs. It was all talk related to game fowl. The government was looking to put me behind bars for four years and a one-year supervised release. That was outrageous to my wife and me. The trial judge granted a variance for one year behind bars and three years of supervision, because there was no animal cruelty. He said that this was not going to ruin my life. I am now a convicted felon. I have never had a criminal record, other than a misdemeanor DWI charge years ago. The judge said I needed to get a new hobby. In a previous hearing, he said, "You really could not tell what Mr. Carrano was doing, but a jury found him guilty." How a farm case could go to trial in the southern district of New York remains a mystery.

When Gina and I first had our farm raided, we went through indescribable emotions—anxiety and disbelief at what had happened to our lives over Facebook comments and my acceptance of four chickens from a GBA member in the Bronx. He received those birds back, minus a rooster that was killed on a farm walk five hours from my farm. The sad fact is that the government knew this but still decided to go ahead with the conspiracy charge. It took them five months to come. They didn't like that I "showed no remorse." I didn't understand what I was supposed to be remorseful of.

The one count of conspiracy was nothing more than a desperate attempt to get a conviction, and it really damaged my life, as well as my wife's. There is not a soul alive that could testify to me ever doing something illegal with a chicken. That is why I could not take a plea deal and had to take it to trial. Our family friends and church family were all beside themselves. God led me through the valley of decisions and told me to stay the course. Righteousness is what this was all about.

To date, we have tens of thousands of people supporting us, for which we are grateful. We have gotten good lawyers now. Things are still difficult at this time, but our faith has stayed strong. Our marriage has proven to be something made from nothing less than the love of God. I believe that I will be the man God has called me to be through all of this.

Printed in the United States
by Baker & Taylor Publisher Services